HI-TECH JOBS WITHOUT COLLEGE

BE A COMPUTER SUPPORT SPECIALIST

by Isabel Teitelbaum

BrightPoint Press
San Diego, CA

BrightPoint Press

an imprint of ReferencePoint Press, Inc.
Printed in the United States

For more information, contact:
BrightPoint Press
PO Box 27779
San Diego, CA 92198
www.BrightPointPress.com

LIBRARY OF CONGRESS CATALOGING-IN-PUBLICATION DATA

Name: Teitelbaum, Isabel, author.
Title: Be a computer support specialist / by Isabel Teitelbaum.
Description: San Diego, CA: BrightPoint Press, 2025 | Series: Hi-tech jobs without college | Audience: Grade 7 to 9 | Includes bibliographical references and index.
Identifiers: ISBN: 9781678209360 (hardcover) | ISBN: 9781678209377 (eBook)
The complete Library of Congress record is available at www.loc.gov.

CONTENTS

AT A GLANCE

- The information technology (IT) field is growing.
- Computer support specialists work in the IT field. They analyze and solve computer problems.
- The best computer support specialists are skilled in communication and problem-solving. They are also self-motivated.
- Becoming a computer support specialist doesn't require a college degree.
- Additional experience and certifications can advance a support specialist's career.
- Internships, apprenticeships, and entry-level jobs are a great way to gain experience. Some of these opportunities allow people to get paid while learning.

- Computer support specialists work in many different places. They work for private businesses, the government, school systems, or themselves.

- Computer support specialists can work remotely from places with internet access.

SUPPORTING COMPUTER USERS

Marie Naraval likes working from home. She can take her time in the morning. She makes breakfast before starting work at 9:00 a.m. She sits down at her computer. She logs into her company's network. This gives her access to the company's system from home.

Naraval is an information technology (IT) professional. She works as a computer support specialist. She joins a daily video

Some IT professionals prefer to work from home. Others enjoy working at the office.

chat meeting with the IT Service Desk. Each member of the team talks about what they're working on. Then she looks at open support tickets. Users create support tickets when they need help with technology. Naraval spends a lot of time helping users.

Some computer support specialists use Voice over Internet Protocol (VoIP) systems, which allow specialists to make phone calls using the internet.

"Some of the common requests are systems access requests. If users need access to certain systems . . . they open a ticket for approval," she explained.[1]

One of Naraval's other tasks is to answer phone calls from users. The phone line is usually not busy. She has time to work on other tasks. She also makes sure to enjoy a treat. One day, she tries a new kind of chocolate. This treat helps her work to the end of the day.

WHAT DO COMPUTER SUPPORT SPECIALISTS DO?

In a world full of technology, computer support specialists play a big role. Specialists work with computer networks. They **troubleshoot** issues. IT specialists

Not everyone who uses computers in daily life knows how they work. Computer support specialists help these people resolve issues with technology.

such as Naraval are essential to many industries. Schools rely on computers. So does the government. Most businesses do, too.

The internet is a big part of most people's daily lives. People work online. They bank and shop online. This means that companies need to have reliable technology. Computer support specialists help keep computer networks working smoothly.

CHAPTER ONE

EXPLORING COMPUTER SUPPORT

Technology has made people's lives easier. People depend on their devices. But sometimes people need help with the technology they use. There is a need for people who know how to work with technology. This is the role of computer support specialists. These workers are also known as technical support specialists.

The need for these specialists is rising. The number of jobs in this field is

Some people use their smartphones for business purposes. Support specialists may need to help these users with their devices.

expected to rise by 5 percent from 2022 to 2032. That is faster than the growth rate of the average job in the United States. New technology will be developed in the future. Support specialists will help people use it.

COMPUTER SUPPORT IN THE MODERN DAY

Support specialists must stay up to date with new technology. One of their main tasks is solving computer issues. They also maintain devices. They work with **hardware** and **software**. They help secure computer networks as well. A computer network is a set of connected computers. Companies use networks to share data. Support specialists test networks to make sure they're working.

Computer networks may use servers, which are computers that store or process data.

Support specialists usually work under network **administrators**. These are the people in charge of the network.

Support specialists solve all kinds of problems. One common task is resetting forgotten passwords. Another is recovering lost files. They also help people use software. Support specialists solve issues

Modern hospitals rely on computers to store and process medical data.

with hardware, too. One user's computer may be overheating. Another person's keyboard might stop working.

Lots can go wrong with technology. Specialists begin their day never knowing what to expect. Felicia Thomas is a computer support specialist at a hospital. She works with many kinds of devices. "We touch everything in the hospital—even the cafeteria cash registers," she joked.[2] She begins her day by checking her email. Usually there are a few tickets waiting for her. She puts them in order based on urgency. Then she begins to work. She might need to help a new employee log into

How Many Devices Do People Use?

Globally, the average person had at least three electronic devices in 2023. The most common devices were cell phones, laptops, and desktop computers. Computer support specialists help keep these devices working.

their account. Or she might fix a broken computer. Most issues that Thomas solves are simple. But sometimes she needs to fix a big problem. Some tasks can take almost an entire day to solve. At the end of the day, she follows up on closed tickets. She makes sure there haven't been any more issues.

A REWARDING CAREER

Being a specialist comes with perks. One is being able to work in many industries. Another perk is job variety. Specialists can fill several roles. They can choose the one they like. Two of the most common entry-level positions are help desk technician and IT support specialist. The help desk usually deals with

Help desk technicians may help users within their company or users outside the company, including customers.

Becoming a computer support specialist does not require a college degree. But it usually requires some studying.

simple problems. IT support often deals with more complex issues.

Computer support specialists enjoy a competitive starting salary. They are in high demand. The median yearly pay for support specialists in May 2022 was $68,050. Being a specialist is an important but demanding career. Most entry-level jobs require training or certification.

CHAPTER TWO

TRAINING FOR THE JOB

Being a computer support specialist does not require a 4-year college degree. But most companies do ask for a high school diploma. Passing General Educational Development (GED) tests can stand in for a diploma.

Math and science classes help prepare a student for a technology career. Students should also take computer science classes. Many high schools

Many high schools have computer labs. Some students discover a passion for technology in these labs.

offer such classes. Penn High School of Mishawaka, Indiana, takes things one step further. It offers a computer science internship. This program teaches students to repair electronic devices. They fix about 4,000 laptops each year.

Aidan Kaczanowski was an intern at Penn's program. He said, "I've learned a lot about troubleshooting and fixing computers." But the program taught him other skills, too. He explained that he learned how to talk to customers. He said, "I've got a lot of experience with customers. . . . I talk to them, ask them about their problems, and help them figure out what to do. My favorite part is when you finally fix it and you have the satisfaction of finally figuring out what the problem is."[3]

Specialists must be able to explain what is wrong with a user's computer in a way the user can understand.

TRAITS FOR SUCCESS

Support specialists benefit from having certain skills. Communication skills are useful when helping users. Specialists must be able to listen to the user's problem. They need to ask the right questions. They must also be able to clearly explain solutions. People who seek computer help often feel frustrated. The support specialist must be patient. IT customer service expert Don Crawley says, "Sympathy and understanding for the user's difficulty is always appropriate, as long as it's **authentic**."[4] Learning multiple languages can allow a specialist to help more users.

Being able to solve problems is a big part of the job, too. Most of a support specialist's day is spent solving computer issues. This can require creative thinking. Attention to detail also helps. This trait can help a specialist find small issues quickly. Fixing small issues can help users avoid larger problems later.

Specialists need to motivate themselves. They need to take responsibility for their own learning. New software and devices

The parts inside computers can fail in countless ways. Figuring out where the problem is can require creativity and critical thinking.

are always being created. Staying up to date with new technology is important.

CERTIFICATIONS AND BOOT CAMPS

Certifications can raise a support specialist's chances of getting hired. A certification is a document that shows that a person knows something. The CompTIA A+ is a common IT certification. It shows that a person knows the basics of IT. People who earn it are ready for many IT jobs. Computer support specialist is one of them.

There are several CompTIA exams. People can choose the skills for which they

Some online courses involve watching live or prerecorded lectures by instructors.

want certification. The exams usually cost a few hundred dollars each. Dakota Snow is an IT support specialist. He got advice when starting his career in IT support. He said, "Everyone recommended that you have at least your A+ certification to get into the field."[5]

Preparing for these exams can be hard work. Many people take online courses or boot camps. These programs usually do not have experience or education requirements. There are many online courses that prepare people for CompTIA exams. Some are given by colleges and universities. For example, Oklahoma State University offers online courses for CompTIA exams. Students have access to course materials for up to a year. They can complete the courses at their own pace. Online courses can fit into people's busy schedules. Another option to consider is boot camps. Boot camps are short, in-depth training programs. People may have several hours of class and homework each day. However, they also usually take much less time to finish. Oklahoma State University also offers a boot camp for the CompTIA A+ certification. It takes only 5 days to finish.

WORK EXPERIENCE

Many good jobs don't require a college degree. In 2020, 37 percent of recent high school graduates did not enroll in college. There are many reasons for this. Not everyone has the time or money to attend

A carpenter is an example of someone with a trade job.

college. And college is not the only path leading to a good career. It is predicted that 60 percent of new jobs created between 2020 and 2030 will not require a college degree.

Being a computer support specialist is considered a trade job. A trade is a job that requires skills or training typically gained outside of a 4-year degree. These jobs lack the barriers that more advanced roles can have. But that doesn't mean they can't provide good pay and benefits to employees. Training for a trade job is typically less expensive than a college degree. And the path from learning to working is usually shorter.

People can get an IT job without a college degree. But it's still important

IT Without a College Degree

About 25 percent of people in the United States who work in IT do not have a college degree. Frank J. Segarra worked as the CEO of a tech company. He did not have a degree. He said, "You can self-educate, come in at the junior level as an intern and work your way up."

Quoted in "Get into IT without a Degree," CompTIA, *n.d. www.comptia.org.*

People can apply to become an apprentice the same way they apply for other kinds of jobs.

to gain experience. Internships and apprenticeships are good places to start. An internship is a temporary position for people wanting to gain work experience. Internships can be full or part time. They can be paid or unpaid. Sometimes an internship leads to a job offer.

Job candidates with experience as interns or apprentices can make the case to employers that they are ready for the job.

In an apprenticeship, a person learns a job by working under someone with experience. Apprenticeships are usually full-time, paid jobs. They are another great way to gain IT experience. Grace Burdett was an apprentice at the company IBM. She said that doing an apprenticeship helped her figure out what she wanted to do next. There are apprenticeships that train people to become computer support specialists. For example, the company Intequal offers apprenticeships in help desk support and IT support.

Internships and apprenticeships can be great ways to get started at a company. They can make people stronger job candidates. The company will already know about their skills.

CHAPTER THREE

WORKING IN THE FIELD

The daily lives of computer support specialists can vary. Many industries employ support specialists. Some specialists work for businesses. Others work for schools or the government. Some even work for themselves. Support specialists may focus on certain IT issues. For example, hospitals might employ specialists familiar with the health-record program the hospital uses. Other specialists

Some computer support specialists need to travel to various sites in order to do their jobs.

San Jose, California, is a city located in Silicon Valley. Many tech companies are headquartered in the city.

work on broader IT issues. These specialists may work at help desks.

Computer support specialists are employed across the United States. But certain states are hubs for technology. In 2022, California and Texas had the highest computer support specialist employment rates. California's Silicon Valley is known for its technological **innovation**. Texas is also home to technology hubs in Austin, Houston, and Dallas.

Some businesses have their own teams of specialists. These teams can raise a business's **productivity**. Working on such a team can be a great experience for specialists. Most specialists work for big companies. These companies often provide good benefits and pay. They can also provide more opportunities for advancement.

Another path is to work for a Managed Service Provider (MSP). An MSP is a company that provides IT support for other companies. These jobs can be more demanding than working for a company's in-house IT department. The work is more fast-paced. Most issues that an MSP deals with are solved within 30 minutes. An MSP can be a good place to find an entry-level job and learn quickly.

Vincent Humble started his career with no college education in IT. His first job was at a help desk for a big company. He mostly helped people with simple issues. It taught Humble a lot about computers. He said,

Because computer support specialists usually work on teams, the ability to work well with team members is an important skill for specialists.

"Once you get into that first job, you start getting experience. And experience is the most valuable thing you can have to continue on in this field."[6]

Humble liked the low-pressure nature of his job. He used the job's slower pace to learn more. He explored other software. He was eager to prove himself to his company. He asked his boss what he should learn next.

HELPING PEOPLE REMOTELY

Some support specialists work in an office. Others work remotely. There are benefits to both. Being in the office can help employees work together. Working remotely allows people to choose where they work. This can save them time and money spent

Many people who work from home benefit from setting up a home office from which to work.

on traveling. Working remotely can also improve one's work–life balance. Remote jobs can give people more flexible hours. People can choose hours that work for them. This gives people more freedom in how they spend their time. People should figure out for themselves how they like to work. Then they can look for jobs that fit what they want.

Patrick Cuneta's first job was selling phones at AT&T. Customers would ask him for help with their phones. He learned how to help people with technology. Cuneta liked working with customers. He also liked giving technical support. He decided he wanted a new job. He would become a tech support specialist. He didn't have a college degree. But he did have a basic IT certification. He began to apply for jobs. Soon, he had an offer. He trained for the job in person for a few weeks. After that, he began to work remotely. He said, "I absolutely love working from home. . . . I definitely thrive in this kind of setting."[7]

Some support specialists help users remotely over the phone or online. This lets them help people all over the globe.

Some support specialists work a night shift, which means they work at night and have time off during the day.

Some businesses offer tech support around the clock every day of the year. It is common for problems to come up outside normal working hours. Around-the-clock support means that users can have a problem fixed before the next business day. Specialists in this role must be flexible. They may need to work nights or weekends.

SUPPORT TOOLS

Support specialists rely on several software tools to do their job. One tool is a help desk ticketing system. This is the software that helps specialists keep track of tickets. Another tool is live chat software. It allows specialists to send messages to users in real time. This software can help when specialists and users are in different places.

Support specialists also use physical tools. Screwdrivers can help when fixing a computer. A flashlight can help too. Sometimes the problem is with a cable. A cable testing kit can help a specialist figure out the issue. Cleaning supplies are also important. A computer can slow down if dust builds up inside.

One reason computers get dusty inside is that cooling fans suck in dust.

There are tools that support specialists can use to help users remotely. Users can grant a specialist remote access to their computer. This is most often done with a program the user downloads. This program allows the specialist to access the user's computer. This way, specialists can fix issues themselves.

The Best IT Tool

Zach Hill has had several jobs in technology across a career lasting more than 20 years. He talked about how he used search engines at work. He said, "[A] big tool or resource that I use . . . is the most obvious answer . . . Google. Google is the best tool for everything."

IT Career Questions, "How to Stay up to Date in Information Technology - Learn More about I.T.," YouTube, *June 23, 2016. www.youtube.com.*

CHAPTER FOUR

LOOKING AHEAD

In the future, there will be new problems for support specialists to tackle. These could include cybersecurity threats. Important data is kept in computer networks. Cybercriminals often use technology for their own gain. People need to know that their data is safe. This is very important for sensitive data such as medical or financial records. Support specialists will help people protect their data.

When people shop online, they share financial information such as credit card numbers. Computer support specialists help protect this information.

F2
F3
F4
F5
F6
Caps Lock
Shift
Fn
Tab

Stolen passwords are a common cybersecurity threat.

Cybersecurity threats are typically handled by cybersecurity specialists. But computer support specialists play a role too. The help desk may be the first place a user goes after an incident. For example, a user might receive a suspicious email. They might not know what to do with it. So they take it to the help desk.

The specialist at the desk needs to be able to tell whether the email is a threat. If it is a threat, the specialist might contact the company's cybersecurity department.

Support specialists also need to be careful themselves. Sometimes criminals target help desks. Specialists at the help desk may have information the criminal wants. In 2023, criminals targeted the company MGM Resorts. They pretended to be users and called help desks. They convinced help desk members to give them access to the system. Then they stole customer data. The company had to shut down parts of its network. This cost the company money. Following safety rules can keep specialists from making costly mistakes.

EMERGING TECHNOLOGY AND TRENDS

Artificial intelligence (AI) is an emerging technology. Its uses may become widespread as it improves. Specialists will need to know how to help people use AI. Support specialists may need to learn how to use AI themselves. AI could help specialists sort tickets by urgency. Tech writer Madhu Murali says that this could "help . . . agents focus their efforts where

AI in IT

Companies worldwide are using AI in their IT departments. AI can detect security threats. It can also provide customer service. Companies such as Amazon and T-Mobile offer customer service through AI chatbots.

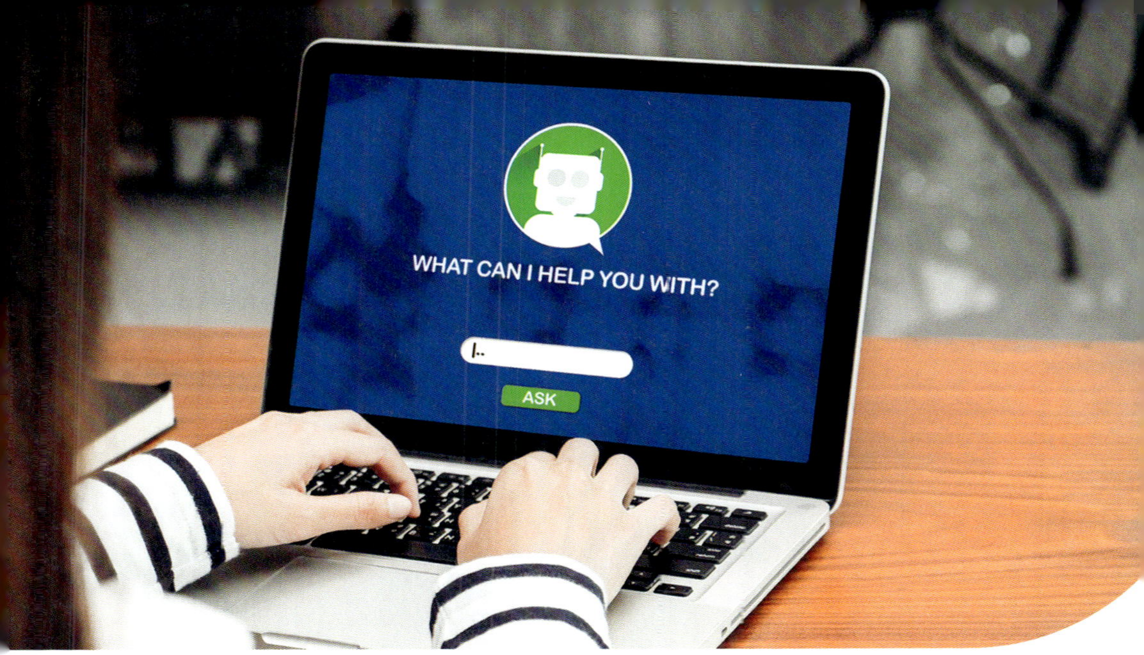

Because AI chatbots can help multiple users at the same time, many companies are taking advantage of this technology.

needed and improve the efficiency of . . . help desk[s]."[8]

A big challenge facing IT is that employers want support specialists to have more skills. Specialists will have to grow as technology does. This could mean earning more certifications. It could also mean learning to use new software. Getting more experience can help specialists advance in their careers.

ANNUAL EMPLOYMENT OF COMPUTER SUPPORT SPECIALISTS

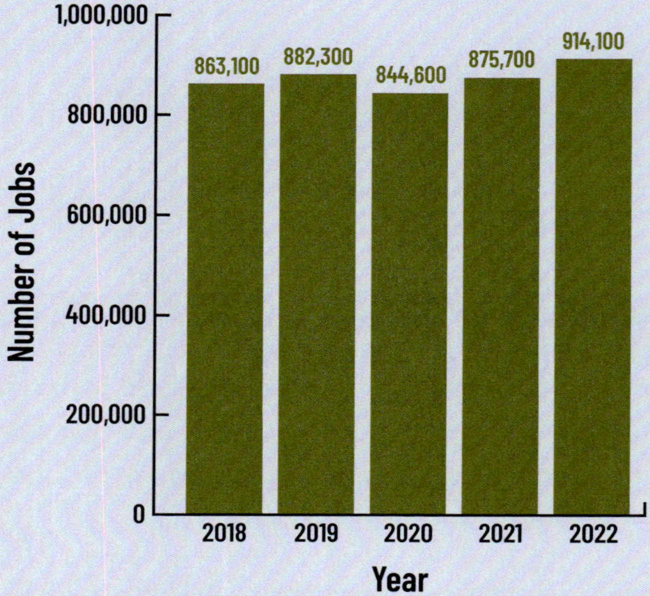

Source: "Computer Support Specialists," US Bureau of Labor Statistics, *September 6, 2023. www.bls.gov.*

This bar graph displays data from the US government on the total number of computer support specialists employed each year between 2018 and 2022.

PERKS OF BEING A COMPUTER SUPPORT SPECIALIST

Computer support specialists are in demand. Gaining experience as a specialist can lead to higher positions.

System administrator and computer technician are two such positions. Sometimes being a specialist can lead to jobs in related fields. Some specialists become software engineers or cybersecurity specialists. Computer programmer and web developer are other options. Experienced specialists can also become self-employed. This can give people more control over their schedules. It also allows them to choose which jobs they accept.

Technology will continue to change how people live. Sometimes world events can change how people use technology. The COVID-19 pandemic reminded people how important it is to have reliable technology. Many people worked and studied online

People relied on video-calling technology during the COVID-19 pandemic. Some specialists needed to learn how to help users with this technology.

during the pandemic. This shift had a lasting impact. In 2022, about 60 percent of organizations that expanded their remote-work options planned to keep them for good. On average, people across all careers in the United States worked remotely 4 days a month in 2023.

It is estimated that 32.6 million Americans will work remotely by 2025. Almost one out of every six companies works without an office space. The push toward remote work is a trend that many workers support. This trend will cause people to be even more dependent on their devices. Support specialists will help set up and maintain this technology.

There are many reasons to become a computer support specialist. Helping people solve problems is rewarding. So is gaining a deep understanding of the latest technology. The demand for specialists means there will be lots of opportunities in the future.

GLOSSARY

administrators

people who manage and direct a system or organization

authentic

genuine, true, or real

hardware

the physical parts of electronic devices

innovation

the process of introducing a new idea or advancement

productivity

the level of efficiency of a process

software

the programs that a computer uses

troubleshoot

to identify problems in an electronic device

SOURCE NOTES

INTRODUCTION: SUPPORTING COMPUTER USERS

1. East Charmer, "A Day in a Life of IT Support | Working from Home," *YouTube*, November 18, 2021. www.youtube.com.

CHAPTER ONE: EXPLORING COMPUTER SUPPORT

2. Quoted in India Ogazi, "A Day in the Life of an IS Support Specialist," *Impact* (blog), *UTMB Health*, April 21, 2015. www.utmb.edu.

CHAPTER TWO: TRAINING FOR THE JOB

3. Quoted in Penn-Harris-Madison School Corporation, "Excellence Happens Here: Penn High School Computer Science Internship," *YouTube*, February 11, 2019. www.youtube.com.

4. Don Crawley, Author of the Compassionate Geek, "The Six Steps in a Successful Tech Support Session: Customer Service Training 101," *YouTube*, April 8, 2013. www.youtube.com.

5. The Bearded I.T. Dad, "My Path to Becoming a Self-Taught I.T. Support Specialist," *YouTube*, March 6, 2019. www.youtube.com.

CHAPTER THREE: WORKING IN THE FIELD

6. Vincent Humble, "A Day in the Life of IT Helpdesk Support," *YouTube*, March 18, 2023. www.youtube.com.

7. Patrick Cuneta, "How I Got a Remote Job in IT (Tech Support/Help Desk)," *YouTube*, September 17, 2021. www.youtube.com.

CHAPTER FOUR: LOOKING AHEAD

8. Madhu Murali, "9 Ways You Can Use AI to Improve Help Desk Management," *Service* (blog), *HubSpot*, November 6, 2023. https://blog.hubspot.com.

FOR FURTHER RESEARCH

BOOKS

B. Keith Davidson, *IT Technician*. New York: Crabtree, 2022.

Cynthia Kennedy Henzel, *Be a Cybersecurity Specialist*. San Diego, CA: BrightPoint Press, 2025.

William Potter, *Computers*. New York: Gareth Stevens, 2022.

INTERNET SOURCES

"Computer Skills for Kids," *Maryville University*, n.d. https://online.maryville.edu.

Livia Gershon, "What Is Information Technology (IT)?," *Southern New Hampshire University*, August 11, 2022. www.snhu.edu.

Justin Shook, "What Are the Roles of a Computer Support Specialist?," *Southern Careers Institute*, May 23, 2022. www.scitexas.edu.

WEBSITES

ApprenticeshipUSA
www.apprenticeship.gov

ApprenticeshipUSA provides information about apprenticeship opportunities in a number of fields, including information technology.

CompTIA
www.comptia.org

CompTIA offers a variety of certifications for people looking to become computer support specialists.

US Bureau of Labor Statistics
www.bls.gov

The US Bureau of Labor Statistics provides data about many careers in the United States, including computer support specialists.

INDEX

IMAGE CREDITS

Cover: © AS-Artmedia/Shutterstock Images
5: © PeopleImages.com-Yuri A/Shutterstock Images
7: © RossHelen/Shutterstock Images
8: © chainarong06/Shutterstock Images
10: © Phovoir/Shutterstock Images
13: © goodluz/Shutterstock Images
15: © dotshock/Shutterstock Images
16: © lightwavemedia/Shutterstock Images
19: © fizkes/Shutterstock Images
20: © dotshock/Shutterstock Images
23: © Tyler Olson/Shutterstock Images
25: © fizkes/Shutterstock Images
27: © Africa Studio/Shutterstock Images
28: © Chay_Tee/Shutterstock Images
31: © mavo/Shutterstock Images
33: © fizkes/Shutterstock Images
34: © Viktoriia Hnatiuk/Shutterstock Images
37: © Olinchuk/Shutterstock Images
38: © stellamc/Shutterstock Images
40: © PeopleImages.com-Yuri A/Shutterstock Images
42: © fizkes/Shutterstock Images
44: © Sfio Cracho/Shutterstock Images
46: © thcdonal88/Shutterstock Images
49: © Kartinkin77/Shutterstock Images
50: © Iurii Stepanov/Shutterstock Images
53: © Wee Dezign/Shutterstock Images
54: © Red Line Editorial
56: © fizkes/Shutterstock Images

ABOUT THE AUTHOR

Isabel Teitelbaum is a Minnesotan author who grew up on Lake Superior. She writes poetry, short stories, and children's books. She believes in the power of the written word and in engaging readers from a young age. Growing up, she found inspiration in literature and hopes to promote the same love for reading in others.